PLATO'S LADDER

PLATO'S
LADDER

Stephen Romer

Oxford   New York

OXFORD UNIVERSITY PRESS

1992

Oxford University Press, Walton Street, Oxford OX2 6DP
Oxford New York Toronto
Delhi Bombay Calcutta Madras Karachi
Petaling Jaya Singapore Hong Kong Tokyo
Nairobi Dar es Salaam Cape Town
Melbourne Auckland
and associated companies in
Berlin Ibadan

Oxford is a trade mark of Oxford University Press

First published in Oxford Poets
as an Oxford University Press paperback 1992

British Library Cataloguing in Publication Data
Data available

Library of Congress Cataloging in Publication Data
Romer, Stephen.
Plato's Ladder / Stephen Romer.
p.   cm.—(Oxford poets)
I. Title.   II. Series.
PR6068.046P57   1992   821'.914—dc20   92-7804
ISBN 0-19-282986-6

Typeset by J&L Composition Ltd, Filey, North Yorkshire
Printed in Hong Kong

CONTENTS

I

II

III

ACKNOWLEDGEMENTS

Grateful acknowledgement is made to the editors of the follow-
ing journals and magazines in which many of these poems first
appeared: *Kalejdeskop* (Łódź), *New Statesman*, *Numbers*, *Poetry
Book Society Christmas Supplement 1987*, *PN Review*, *Poetry Review*,
Smale Fowles, *Times Literary Supplement*, *Yale Review*.

'Adult Single', 'Picardy', 'Plato's Ladder', 'Roissy', 'The
Weight of It', 'The Work', and 'Virgin' were broadcast on Radio
3's Poet of the Month series in December 1989.

The author wishes to thank the British Council and the
Institute of English Philology of Łódź University for their joint
support throughout 1989 to 1990.

I

'And the true order of going, or being led by another, to the things of love, is to begin from the beauties of earth and mount upwards for the sake of that other beauty, using these as steps only ...'

Plato, *The Symposium*

On the Golden Isle, Amboise

There's a spring-green boat
moored along the Loire,
waterlogged, unmanned,
dreaming in a green bend
of harbouring river . . .

In a grove of aspens,
a haze of mirrors,
a brushwork of young oaks
with tender leaves
and a chaffinch

you need not think
of violent deaths
now or as they were
hanged like fruit
from the Castle balcony

but of a gentler kind
your back against the planks
eyes up to the green
into which you disappear
sundazzle and shadow

where the little skiff
imperceptibly loosened
grazes the sandy isles
of the braided river
nosing westward to the sea.

Adult Single

As if a diary redeemed the time
I bring it up to date in a solemn

trivial rite, as if the recent past
were mastered like the latest

headlines, as if, once and for all,
I could get things under control

by jotting them down on this hurtling train
and disembark, born again

—but the appetite dies for reforming prose ...
Papers fall from our hands, my neighbours

and I succumb
to warmth, myopia, constant rhythm

where fog has blotted out the landscape.
Cradled to sleep

I'm settling like contents in transit,
my head slips further down the seat,

my thought into solipsism, a sealed shell
of privacies, lulled from level

to level down a fault in the ocean floor.
I'm travelling back, through millennia

of evolution to the whiskered fish,
on the black stream of my single wish

to linger in perpetual motion
behind blue windows and the lash of rain.

Work

When shall we ever begin?
Swept mercilessly clean

there's a billion billion stars
in the skylight, and our chairs

make their strict companionable arc
with the fire. We're ready for work,

it's the moment we've been waiting for.
After a day of trial and error,

triumph and tantrum,
our baby's down and milky calm.

Our stubborn house
is nearly at peace,

as tamed as it ever will be.
We're dosed with tobacco and Irish tea,

there's a wall of books to be read,
hours of encircled lamplight ahead,

cimmerian voices crying to come in.
Now if ever we might begin

when the cold is eerie but it's warm in here
and we sit without moving, drunk on the idea.

The Work

Settling to work is like the idea
of Venice, a cutglass Serenissima

flashing in the mind's eye;
or a tract of country

we left unexplored one summer,
the small hills winding away

into the memory of what was to come,
the peace and space and time

we promised ourselves, but drove on
gazing back at the work to be done

receding in the rearview mirror.
The Work itself is always there,

like Ithaca, grown lustrous in postponement,
or Penelope, or what we most want

and best avoid, our tacit destination.
There are passages in Titian

of lapis lazuli, distances
at the end of experience,

cool water, bridges, hills, a landscape
of reconcilement and essential time

where the work might take shape
under our hands, if we could get there,

and recover those we have lost,
from the first to the last,

if we could say *the last*, renouncing
such kisses as may come

or places we never saw, the idea
of Venice among them, banished forever

from our high blue peak of concentration.

The Inimitable

'. . . the sketch that is our life is a sketch for nothing . . .'
—Milan Kundera

And yet, there are lives in outline
—not that discouraging skeleton

of facts, the mournful *curriculum vitae*,
but the dash of personality,

'an unbroken series of successful gestures',
the caressing shower of Gatsby's shirts

or the transience
of Katarina Witt on the ice,

the immaculate figurework
of a human swan, a white ecstatic,

her calligraphy of perfect control.
The bubbles rising in his crystal bowl

elicited the wit
of Mallarmé at his banquet

murmuring his epitaph, *rien, cette écume*,
serenely blowing at the spume

of his own speculative arctic.
The few pages, the prodigal talk,

the father, friend, husband
compose the diamond

of the life of the man
into a single arc of attention.

Matisse gave all his will
to growing fluent with a pencil,

and through lethean satieties
from odalisque to odalisque

drew swiftly on, till he too disappears
among stripy fish and swimming birds

in the Polynesia of his own invention.

The Other House

(*in memoriam John Strevens, 1902–1990*)

To our discomfort, he spoke near the end
of another house, suspended within this—
an airy hospitable place
where his daughters,
all Cordelias now, and spared,
stretched their arms towards him.

How he lived there was not clear,
nor who we were either,—
and there were other faces
in the curtains and the bedlinen,
gargoyle and demon
staring back at him . . .

He painted lighted flowers, uncomplicated
smiling children,
but had his emblems too—
violin and bow, the dandy's cravate,
a deathmask of a girl, the *Inconnue*,
pulled from the Seine, angelic serene,

the kind of thing
adored by ruined Dowson
—the drapes, the lamp, the pallor—
those cadences
of lifelong valediction
closing in a dream.

A tiny lilypond
was sunk in his honour
to lure him like late Monet
—he refused the role—
damn boring, open air,
and shuffled on down

hat and stick
to the wooden studio
of miraculous imagination
where the brilliance of Sargent
shone serene as Renoir
if he sat quite still

and let the palette swirl
your darks your lights
applegreen and cadmium
rapidity, panache,—his style
had the flourish
of a full-blown peony

and when he died, what he knew
went up with him,
being curious, wide, wayward,
and all his own; there were traces,
scuffs of paint on a graphite Mozart,
canvases and books.

Among them, something called
The Christian Agnostic, whatever that is;
but when he floated free
after the years of shuffle and shrinkage
may it have been to some such vestibule,
humane, fragrant, his other house.

Beneath the Tree

(for Douglas Oliver)

The shaken oak is a vessel
of lights, and they do not spill,

held in a cellular system
of blossoming and extinction;

we perceive it simultaneously,
and beyond it the sky,

with its breadth and curve and tenor,
a brilliance at the edges of the retina,

the natural backdrop for vision
to take shape, step forward and shine

as the single accomplished aggregate
of all we have felt and thought

unshareably about a face
we have loved; but it swims out of focus,

its lines and colours will not hold fast,
and the sky itself is an iconoclast

perpetually sweeping the canvas clean.
If thought could travel out and back again

and not fall crooked, eclipsed or predetermined;
if we could hold music in the mind

and receive, as something solid,
the *Blessing of God in Solitude*

or the *Fountains at the Villa d'Este*,
—but the trills and octaves stream away

to leave us asking what it was, exactly,
which flashed through the cerebral tree

of nerve-cell and synapse, whatever they are,
imponderable matter

where all is registered, whether a conversation
or a lifetime of living with someone

however right it goes, or wrong,
we have the same possessive longing

to see it entirely as it never is,
cohering like an oak tree in repose.

Cucaratxa

The present is another country
belonging to children, or the masked fox
zigzagging nose down on the mountain.

What must it take
to compel me into it;
an earthquake

or a herd of deer
stepping briskly down a vertical?
The close, unhurried scrutiny

of nine Egyptian vultures
wheeling out
from a rockface?

The present is the prayer
of the mantis
to paralyse the fly

and the cellular eyes of the fly
are riveted on the present . . .
There's an aspect of hell—

the protean mind fixed
in a circular system
of its own devising,

sealed from the here and now
in the single shape
of its own despair . . .

For half an hour
I would inhabit
the greengold body of the grasshopper,

chafing my red hydraulics
for the sake of it,
or punting for dear life in the stream,

sluggish in shadow, galvanized in sun,
chirpy forgetfulness
my only store.

Mythologies

1 Temptation

What did the lake god say
in his annunciation
—pulsings of his turquoise body
through the blackness of a pine—

or what did I ascribe to him?
Apollonian light,
mineral tenacities.
Could he not excoriate

those clinging tendrils
that persist in me
of merely organic fibre,
fear, obsession, desire, envy?

Bind me up with granite,
the sapphire gentian at my feet.

2 Painter

(for Caroline)

You are my Proserpine of summer
knee-deep in the scabious and mallow,

red-eyed from looking in the light,
as if there were grief or fever
in your exacting tribute
to momentary outcrops of yellow,

painting against the time

when rapidly over the rim
of the granite mountain circus

or in the sealed cranium
at noon on the clearest day,

streams a vocabulary of blackness
no sunlight in the lap of the valley
holds at bay.

3 *Virgin*

We're seldom what we think we are.

Moving out on the printless sand,
self-possessed, a girl who knows her mind—

Phaedra had a mother and a father.

The scrolls of her hem,
the scrolls of the foam,

cry out for an interpreter

but the meddlers on their tripod of cloud
prize that unusual blood
which ached for a touch of the bull . . .

She's moving out, forearmed, determined,
chaste as Artemis, untouchable;

the present belongs to her
(and the past takes care of her future).

The Master in Italy

His thought is chaste as a eunuch
though not unstirred by the Baroque

Ecclesia di Gesù, a lusty baritone
and an opulent young nun.

Indiscretions! To walk in Rome
without an overcoat, and countenance boredom

as an elegance in the young,
though it isn't quite done

to preen in bar mirrors or recline
in the Pincio for a whole afternoon.

Supreme badinage . . . But leave him in the evergreen
fatigue of a dusty villa garden

and he rinses his reader
of approximate despair

by weighing the cornices
of actual light, and speaking sentences.

Bellagio

'It was better even than being in a novel—this being, this fairly
wallowing in a libretto.'—Henry James

1

I'm in deep water, alone with wraiths
and waiters, in a small hotel;
somehow solitude needs justifying
when there's a bed for two, a balcony,
a loveseat and a view of Lake Como,
as if it were all to begin again . . .

Here the Alps *break down romantically*
in other words than mine, but hold us
on the shores of a glacial basin,
where novels of immoral tendency
were conceived by *inflamed young gentlemen;*
and this is a lake of echoes.

Sunset was vermilion, piled
with the cloudflakes of September,
an Assumption housed in a cupola,
the white legs of the Virgin
disappearing into glory . . .
I would lull myself to forgetfulness

or aspire to the Absolute,
and made a start on the journey up
by leaning back in Parma
to see the massed angels of Correggio,
a circus of flesh and drapery,
spiral through the gloom.

Later, in a museum resembling
a labyrinth, hallucinating,
overwrought, I wandered fathoms down,
ready to do something foolish
like fall in love with a painting
and revive an indestructible desire.

2

Leonardo is much to blame
for fixing the equivocal
in its maddening incarnation;
everything promised, everything denied,
and everything negotiable! . . .

His *testa di fanciulla*
undid me in Parma;
being susceptible and rotted
with idolatry and *de l'amour*,
I attached it to a living model

by way of Stendhal
who found the *Belle Ferronière*
resembled his Contessa,
went mad and fled, *to the end of hell*,
down the San Gottardo tunnel.

3

Below the sculpted hills, a traghetto
plies a diagonal over the lake
as I watch from the shoreline, and recall
us here together, two years before,
imagining Liszt and his runaway countess
in their loggia by the villa Melzi.
—We revelled in the l's and g's and i's
of nostalgia, and grew nostalgic
for each other, and for a spellbound past
which our eternal adult talk
could do nothing to revive, as if words
could increase our active tenderness one jot.
There was the black labyrinth of our village
on the lake, the bad food in the airless
dining room, and coming near to madness
on my pillow that night, with the knowledge
that mutual memory was not enough
to hold us anymore; and my hurtful vision
that radiance was all elsewhere,
beckoning through a locked iron gateway,
or fixed perpetually in a frame . . .

And what if the whole tender thing
happened to us again, could we bear it
rigid with our memories and misgiving?
Could we pass through the seven stages
and the two crystallizations
to the place where another's existence
becomes the chief mystery—
how is he or she *possible*?
Could we move to the new music
or learn a fresh vocabulary?
We attempt to gather ourselves up,
all at once, in a single gesture,
as if to say *we have come this far,*
this is what we've learned and what we are
—and shall go no further, having seen enough.
With a careful blend of kindliness,
knowingness and fear, we dissuade
those who might tempt us into life . . .
We'd settle back, but there's no complete escape
from the past trailing behind us
as it merges with the needs of the future.

4

The Signora's tired, the season over;
a light mist breathes on the lake this morning;
the other guests left in a Mercedes.
I'd grasp the extent of my delusion . . .
It's all new money here and 'leisure activities',
and money is the great decider.

With a lot of patience, the mind
and heart can heal. And there is knowledge,
though most of what I learn
I leave behind. This cooling mist
revives the memory of autumn
and whets the appetite for home.

II

'Utterly vain is, alas! this attempt at the Absolute,—wholly!'
A. H. Clough, *Amours de Voyage*

Cautionary Tale

In contemplation at the café:
a dazed young man out of Cavafy,

'the poet nearing his twenty-fifth year'
gone from bankrupt to millionaire

in the small white hours of eros.
He walks into a morning of promise

and sunlight, with a world-including smile,
startled to find business as usual,

and on his body the surplus greatcoat
from the hopeless months, with the torn pocket

where coins disappeared in the lining;
the same body too, not hunching along

this morning though, but cock of the walk,
the lover tolerant of discourteous traffic,

exuding the benevolence
of a man whose work has been praised, once,

in the press ... Now he's at his zenith
in the café; but the waiter's brisk cloth

is applied to the next door table,
change is ceaseless and imperceptible,

the world is moving round him in his dream,
but the world is moving on without him.

Portrait: Premature Mid-Life

By now, he has no
single destination, only
a precarious alternative
with time running out . . .
The early choice
of thrilling guilt
or consoling virtue
has been replaced
by two flaming angels
of recrimination.

He is skinnier
than ever before
and the owner
of several plastic bags
each containing
a forgotten toothbrush.
His books are stored
strategically, according
to unpredictable
laws of access . . .

He spends hours
in traffic jams,
travelling to a car park
on the hard shoulder
of a suburb, to escape
the agents of bad luck
who are even now
fining his windscreen.
They provide an excuse
to leave quickly in the morning.

He is often in cinemas
and confuses
Liaisons dangereuses with
Fatal Attraction;
he devours novels

none of which quite apply
to his case . . .
He decides to keep in trim
by reading deconstructive
revelries over the void.

He is importuned
by certain phrases—
Moral Choice or *The New Life*,
The Children or *The Stock of Memory*;
he rehearses them
in the metro
travelling towards evening,
stirred by perfume
and a frank sexuality
pouting on every wall.

Five o'clock
is the blessed hour
after work and before
decisions, the delicate
negotiation
on a public phone.
The first drink.
He walks decidedly
to the *Café de l'Espérance*
elected

among other things
for the waitress.
He comes alongside
the gleaming chrome
and asks her
about the tulips.
De la Porte des Lilas . . .
He recalls it,
a fragrant name
from a remote life.

Departures

Possibilities are everywhere, only
access to them barred: the great white cruisers
jostle behind grilles at Marseilles,

a fantail of escapes! Or the twinkling
Palma ferry at dock, one sweating night
in Barcelona, where I saw a sign

Bodega FIRMIN licoreria; His Majesty's
sockless Consul would have smiled—
that was a future out of Mexico . . .

I see him whitefaced, in wraparound shades,
sidling into the cool galleries
of the romanesque, to contemplate his own flesh

forked apart by devils
beneath a pair of suave Byzantine eyes.

Serenissima

A mauve *trauergondel* off San Michele,
Venice ankledeep in rain ...

Ezra's laurel and the small graves of children
bore the flood with patience
on the rotting island of the dead.

We came to grief
in the Campo degli Incurabili
and almost to blows on the Bridge of Hell

where you strode ahead,
magnificent in your long red raincoat
impatient for assumption, or just to fly away ...

I decided my only hope
in the universe that day
was the emerald green and cobalt
in the lower pane of a window,
Ss Giovanni e Paolo.

Maltese Fireworks

(*for Peter Vernon*)

Every field a crop of stone,
every post a birdtrap.

On some, the soul in purgatory
peeps from a bowl of flame.

Men grown trigger-happy
from blazing back

at firebombs and the sun,
dawn, noon, dusk and night,

north of Tripoli, east of Sousse,
send up cloudlets on which she rides

Ave Regina Coelorum!
In a barrage of noise

Our Lady of Gunpowder
blooms on the horizon,

a brief red oval
in a circlet of stars.

On the Shiants

(for Adam Nicolson)

Feeding, childrearing, territory:
the murdering blackback circles

but her woolly fledgling
niched among stones

has only colour and silence
to protect him.

Monogamous puffins creak in burrows,
one life, one love

which is where the complications start . . .
We lie out on the stones and talk,

while our sons play spaceships
on sedimentary rock.

There's an ammonite in your hand,
casting gloom on our species:

the brute fact is, we're a form of carbon life
slowly evolving round a G-type star,

—a chastening view,
as we approach a cusp or fold or crunch—

one life, one catastrophe!
Even the stones we sit on

suffer entropy . . . You are more rapid,
impatient of stasis, with a mind

like the joyful skua
exact and fast. You knew

more than the summer had come to an end.

Olivia on the Hebrides

The pale Atlantic rises from the sill,
a candle shows in the glacial window

and the Hebridean midnight lingers green;
is your mind never still?

Relentless reasoner,
head full of theory,

you'd sort us to a tee . . .
Tonight, we had to put in order

goodness, intelligence, beauty—
I chose goodness—priggish, meaningful—

you intelligence and Adam beauty
(letting the others *come with the chances* . . .)

You spoke of a brilliant structure,
a theory of everything to explain

the vagaries of male behaviour
that allows for our silly Platonics,

our falling for the ultimate face,
fixed forever, until it's over,

as much as for the brutal snapping
of the lobster in the creel.

The talk got on to quantum, eerie
transgressions of universal law

when you dropped like your children
headlong into sleep,

the mew of your bright hypotheses
silenced, while the peat smoked

and the green passages
stayed in the horizon.

Plato's Ladder: a Dialogue

'Now I must ask it, what is virtue,
our strong lost word, what is its hue,

its honour, is it an end or a beginning
for such as I in my sixtieth spring

when the fluff is blowing in my eyes
and the deadweight lifting gives rise

to absurd lightnesses on wings,
to pods, seedclocks, haze, trumpettings,

the dogstooth violet, bluebell, heartsease?
What can I make of these,

or of the gentian,
distilled by this April rain

on the upper slopes, if not a quality
engendered, like virtue, petals of the sea

in the mind of a landlocked people,
an idea of ascent, a principle

of detachment from the local pain of love . . .
To climb for wisdom is to move

from the one beloved to the general
and thence—'
                    '—to your "blessed autumnal

calm". You've tried it before. And when you see
blossom from the untamed cherry tree

detach itself to join her gentian train,
you slip down Plato's ladder unashamed.'

The Weight of It

You come out into the floating garden
of early October, there's a mist on your cheek
and you say it's autumn, what have I done

what must I do? and look back at the house
with its cock-eyed face, which you think you want
to leave, but you're seized by a mysterious

invading reverie and stand quite still,
your footprints tracking back in the dew;
the lawn's a mirror where your thought can't settle,

even your language is lost, suddenly
among the grass and the foliage, in the rippling
water of ash leaves against the sky.

And momentarily you're a medium
these presences will trust and pass through
for they know you well, and you them,

such has been the constancy of living here;
then the vertigo of time sets in,
as if the years were gathered in an hour

and you're still standing, rooted in the place,
grown into this garden as the garden
has grown into you, a solitary witness

who cannot easily withdraw.
Then you may ask the garden a question:
what is it continuing for,

what kind of certainty can it supply
after the years of watching
and tending? But there is no reply.

In the Country

There's no escape this evening
from the silence or the light,

no cover; its intensity slants in
and overruns the old glass
which has wept itself
into pockmarks of distortion.

There's no help for the silhouettes
we've become, looking into rooms
where feelings have run so high
so silently and for so long ...

You would turn your back on it
deliberately, at a loss
in front of such a statement,
but early television is flooded out

by its vividness at your back,
its reiteration of beauty,
and behind it stasis and death.

Picardy

1

Today he named the snow. He is himself
fresh snow, and the baby shoots poking through.

When he looks at the lake
an estuary of light

catches him flush on the face
and borrows his startled breath

to utter itself,
immediate voice, immediate water.

In the musical car
he stares at the winter colours

of Picardy; but the violet scale
and tints of fusco-ferruginous

still wait to enter his vocabulary.
That in time I shall supply,

for I need him as he needs language;
we're perfect companions for a journey.

2

I can't so easily explain
my thirst for ghosts and expiation

which touches him yet goes beyond
anything I would have him learn;

what makes me drag my toddler round
the whimpering forest of Verdun,

or stand together in a rising mist
on Douaumont, to overlook a battleground

of squalor and obedience . . .
Trouble closer home has sent me out

to clench myself in a punitive wind
under a bloodshot sky

where I hope to be forgiven
by Our Lady of the Marne

whose infant warns the enemy
Tu n'iras pas plus loin

as my own tugs my sleeve
to bring me away

from this landscape of memorials . . .

All I would have him learn
are the colours of the field,

mild grey and khaki; the blue horizon.

Succession

A fragment of time in the pure state . . .
I remember it

as a mile of revelation,
rain thick as vodka on the windscreen,

midnight blue above the gable
of the Grand Hotel at Cabourg, a double

rainbow drilling the plough . . .
In the same place, I think of it now,

of how the yellow milk in that horizon
is overlaid by this one

rapidly closing down,
and of how my own son

zigzagging through the monochrome
will overlay me in time.

Trust

Estranged in a foreign country,
languageless, how I'm here or why

is baffling; it's carpetless and bare
and I am no one. I've come far

from what I haven't thought of
in so long, the absolute safe love

that held me as a child, when home from term
I'd sit beside the fire in my room,

pyjamas warming on a chair.
Tucked up, the last I'd hear

was the firing boiler tremor . . .
Trust is no less true

than fear,
which comes to look like second nature

yet is not; and nor is dissolution,
for the life's continuous and one

with the child we were
and can recover.

Roissy, December 1987

When the jets make loud and clear
that a couple have lost each other

in the machinery of separation,
and I watch the pain

of the man who's left behind,
what comes to mind

is a fashion still from 1938:
careless of the state

in a prospect of war,
a model highkicks on the Eiffel Tower

in a fantail of silliness;
her dancing shoe makes all the impress

of what is personal
on the riveted steel . . .

And when I witness
the puckered face

of the man left on the ground,
who in a second

will be gone,
driving hard through rain

to loud music and a drench of tears,
I think of that girl on the bars

and of his girl on the plane,
and the lightness of what is human

is heavier to bear
than the thousands of tons of pressure

which rock the Boeing, as it wobbles
on its cluster of wheels

and skitters forward
with implausible speed

as if, by staying motionless,
it would shake itself to pieces.

For Paul

After a long, disturbed absence,
what I recover from seeing you
is not merely speech, but a pattern
of gestures I loved, the regular slight panic
with which you pat all your pockets in turn
to check on money, cards, cigarettes, lighter—
the debonair way you still brush back your hair,
the efficient unfolding of spectacles ...

Maturer? We're older certainly,
greyed, if anything, a touch beyond our years—
warier, not daring, less expectant—
we're not waiting now for any afflatus
or a gladness that came once and not since ...
We share a new admiration for serenity
especially when founded on a desolation—
the sober pleasures of the eighteenth century.

But warming to the calvados, the question
at last—*are you writing?*—the creased brow,
sighs, reserve ... Soon we're back to the favourites,
the few indispensables for when in need—
romantic, intimate, confessional for the most part—
and the old jokes follow, the latest gossip
on the same old group, and all the savour
being with you has, which is of no date.

Petit Air

(for Myriam)

I know, my love, I've preached at you
with all the weary wisdom of Onegin

who froze his Tatyana out
with a plausible commonsense

that sticks in my throat.
I've talked my level best

to turn you away
and hated it—

pleading my past, my age, or worse,
with fingers crossed,

terrified
of convincing you . . .

So now that you've gone
and left me to myself

all I long for
is the baffled candour

in the natural gravity of your face,
that *little air*, which I catch,

—and you catch me catching it—
—bringing your blush to the surface—

prompt radiance
to scatter my better conscience:

bankrupt and disarmed
I draw your smile around me like a charm.

Prophecy

After our ninth separation
I sat in the penitential café

(where a kindly waiter
removed the extra place)

to stupefy myself on wine
and hum *As Time Goes By* too loud

and prepare for the Works of Solitude . . .
Everyone else was a couple,

leaning over little tables
to mould the face opposite

with fingers and palms . . .
I knew I would call you up

in a month or so, and saw
the ragged crenellations

of self-sufficiency dissolve
in the immense dilation

you draw
(as into the mystery of my faith)

over your eyes, veil on veil . . .

At a Glimpse

How often, in Tolstoy,
the cycles of chaos and meaning,
futility and purpose
dejection and joy

hinge on a glimpse of a girl,
a streak of laughter
in yellow, with a black curl
escaping from a kerchief

who is enough to banish
the disabused boredom
the puerile misogyny
the delusion of power

and simply replace them
with herself,
a single high cloud
outlasting the battlesmoke

in the dry eye of Bolkonsky
the wild eye of Rostov
the weak eye of Pierre
the cold eye of Dolohov

before the nihilism
starts again.

Val de Loire

Cantat philomena
sic dulciter . . .

For nights and days
the aspen poplar
has snowed itself

in fluffy miniature
on the lake and woods;
now, it's white underfoot.

At chosen hours
a pair of buzzard
patrols the treetops

and once a kestrel
mobbed the old grey heron
who turned in outraged circles

and creaked to cover.
At Pentecost
two goldfinch scuffled

upside down on lilac
which has rusted since
and the cherry tree

so garlanded and hooped
was stripped to green
in a day.

The jonquils died,
the tulips died,
the iris died,

and now we have
the generous, unkempt peony,
thornless cousin to the rose

with a scent all her own.
This evening
under a white moon

as daylight lasted
a nightingale uttered
his triplet crescendo

heard by Alcuin
downriver at Saint Martin
in his cell,

where he waited
in retirement
for the knock upon the door;

twelve hundred years
his little bird was rapt
from a spray of broom

beside us, quivering.

III

'. . . that fairy-tale defenceless land
on which feed black eagles, hungry
emperors, the Third Reich, and the Third Rome.'
Adam Zagajewski, 'Poems on Poland'

Tablet of the Law

Warsaw, 1990

The people's laureate
strides on his podium,
graduate in Collectivism.

Helpmate Wiesława
uplifted stands parallel.
In spanking harness

they plough the end of history
into a republic of play,
the Palace of Culture behind them

much as in *Babar the King*.
He grips a marble tablet
listing the troika

Marx, Engels, Lenin,
and a fourth, struck out.
The blank aches to be filled:

Friedman? Or Loyola.

The Titans

(for Jerzy Jarniewicz)

The city clings to its grime . . .
Four Titans of Socialist Work
are blacker than most
from implementing the future.

Eyes goggled, ears padded, the men of concrete
are musclebound and cramped
with all their emblems in their arms
bowed beneath a plaster cornice.

On a pristine apron of snow
four bulky moonmen sway and stagger
and stop dead below the stars, drunk
on the weightlessness of a new planet.

Westward-Ho

A spillage of gold light on the sludge
—the dollar shop. Johnnie Walker tips his hat,
smirks and hurries westward off the shelf.
Things are familiar, but not quite right;
there's a cigarette called *Ronhill* and a condom
with a name: *Anti-Baby* . . . A girl in a woolly bonnet
fingers Parisian slips and silks.

In the urinous tunnel to 'Manhattan'
—a clutch of scrapers in downtown Łódź—
the brand new soul of the entrepreneur
traffics in carrier bags and lighters,
the abstruser bits of bits of cars,
Solidarnośč tee-shirts, Catholic literature,
laceless cowboy boots and denim cleaner.

Huddled in the cavernous cinema
I feast on *Wall Street* and consider
the junk bond miracle where greed is good;
I dream of my salary in seven figures,
the hard convertible złoty; and swagger out
like Gordon Gekko, to slither on the street,
brought up short by a wheelbarrow.

Łódź

1

Pedestrian alcoholics
zigzag down the motorway

on the outskirts of Łódź;
there are tractors without tail-lights

and a garish sign
spells *Telimena*, the textile giant

who was once
the aristocratic beauty of Mickiewicz.

I double take the first glimpse
tenements high

made blacker by weak wattage
and cables hanging from housefronts

—it is a gridwork
of unimaginable extension

dissected by a vague arterial
driven through remnants of the ghetto.

There was steam
pouring from every vent and spigot.

2

It was worse than Bleaney's,
the beige flat with green carpeting

salvaged from a very long corridor;
the rooms were a wind tunnel

and the lavatory leaked
above and below.

I went to the *Centrum* for a bite,
was seated at the only unlaid table

in the echoey dining room.
At the far end, a cabaret

was in progress.
A dwarf bicycled on his hands

and a standup comic
shouted like a moustachioed demagogue

at the furniture.
On little plates, acres of egg mayonnaise

stretched away;
and no one came.

3

Nothing could furnish that room,
not my fond mementos,

books, icons, the pocket album
transported wholesale in a trunk.

Simonetta Vespucci
embodies humanist wisdom

but even she, with her coiled hair
and unimpeachable breasts

kept falling from the wall.
Her woman's touch

made no impression
on the uprights of the place.

What remained
was my inert awareness

of decay,
food, skin, paintwork,

and an active neglect,
the deposit of dirt

on my windowsill each morning.

4

The first day, I taught Conrad
or Konrad Korzeniowski

in a windowless amphitheatre
(having groped for a light switch)

to my seminar
of demure beautiful girls.

And this too has been
one of the dark places of the earth

—that meant something,

though later they were vexed
by the difficulties

of taking tea at Howards End.

5

From three o'clock on
at the Grand Hotel . . .

Down on his luck, a haggard Paderewski
was mixing schmaltz

unheard by free marketeers
consulting over vodka

with the vague criminality
that clings to groups of males

talking too earnestly among themselves.
In the pocket cocktail bar

whores and *gintonic*
were on the agenda;

a German in his cups
bargained through the smoke,

slurping and calling out figures,
raising the stakes

with inflationary abandon
until he hit lucky

and the skinny pomaded teenager
became a radiant smile,

trotted over obediently
and leant her head on his heaving shoulder.

Night Life

We splashed out in the *Kaskade*,
a night-club cum insurance building
where a party of Armenians was *en fête*.

A wiry local with a bleeding mouth
was thrown out continually
and reappeared through the first floor window

undamaged and unkillable.
Lambada was the dance that season,
South American athletics

transferred to the swollen North;
the cucumber farmers of Armenia
went at it with gusto.

I scrawled my Christmas greetings
on the back of a paupered Santa
with a Karl Marx beard

sorrowing like Lazarus
at the rich man's gate.
And there we were,

with disco lights, an orchestra
and a bootleg cassette
played over and over

at the open gateway to the west,
licking our dried lips
before an anticipated journey.

Bitter enders
we left with the last drunk
tenderly cajoled by his wife

and pushed by the manager
on to the broken cement
where neon flickered on sludge.

Sweet Confusion

We lay in the spring wind
on a piece of waste ground
outside your block;

inside, your *babcia*,
a tiny martinet with a twinkle,
scoured the flat.

Great creamy clusters
of white lilac
fringed the parking lot

and tulips blew
behind allotment wire.
We walked through silver birches

and found a country restaurant,
with a porsche outside
and a cosy dining room

where we ate roast duck
and drank French wine
and looked across the lawn

to a leafy cottage,
a pocket Cotswolds
for your Manchester of the East.

You were ambivalent
about the new Poland;
detested Russia,

but grew tearful
singing Russian love songs.
I meant to be avuncular, pedagogic,

and read you the Herberts,
George and Zbigniew;
you lent me Joyce,

my bedside book in Polish.
I glimpsed it every night,
my reciprocal good intention.

You developed a need to confess
though as far as I knew
there was nothing to tell.

I even dropped you off
at the red cathedral,
and when it was over

you met me with an exalted smile
at the carrot juice bar.
One month later you were married.

Sailing to Sopot

In a brown smoky room
to the North of Central Europe
we lolled over a tablecloth

and remote waiters
lolled
against the sideboard;

outside, under cloud,
the chrome blue Baltic
sported its swans and oily fringe.

To the right
Gdańsk, with its angular
unsellable gantries

and ahead, the narrow arm
of Hel peninsula
where the dog roses blow.

We never got there, and my feeble pun
to be postmarked Hel
was prevented by the military.

So we went to Sopot,
rounding the corner at Westerplatte
and sailed out of history

on a stretch of open sea
where the Grand Hotel
emerged above the glitter;

it might have been Cabourg,
or the Lido,
with Aschenbach among us

in the little ferry
filled with Polish children.
I could see him, muffled up,

in one of the ample basket chairs,
which faced in all directions
that deserted day in early March

when over a soiled tablecloth
I discoursed on time and the past
so sentimentally you grew restive

and rolled your beer green eyes,
zielono-piwne, longing to be out
in the sound of the sea

szum and *morze*, the words
come back to me, your mouth
was the seashell, close to my ear.

Dissident

A rabbinical poet
with hippy dreadlocks

was king of the clients
on vodka street.

His wife drank brandy
and translated from the German.

He had written himself
into a noble paralysis

nothing I do, dear reader,
will change a thing, not even the poem

I am writing now.
So he sat back, almost supine,

in his chill, voluminous flat
with its frieze and wrought iron balcony,

a leftover from the days
of the textile kings

—breakfast from Berlin,
supper from Paris—

and organized opposition
to the opposition.

Tea for the Native Speaker

Annually, the Professor gave tea
to the 'native speaker'.

She lived in a thirties block
on Narutowicza,

—we called it Narrative Street—
though little went on

besides the sputtering blue
early electrics tramline . . .

A looming, kindly woman
in a narrow flat;

I shuffled over the parquet
on a pair of giant floorpolishers

and sat below the Sacred Heart.
Romola, which I haven't read,

was on the table:
we talked about George Eliot,

awkwardly, and Italy
which was easier. Henry James.

Strawberry Jam. *We Poles are famous
for our homemade preserves;*

and then politics.
We are very worried, you see,

and finally, the criminal abandonment
of '39.

Before I left, she made me a present
of what I supposed a Christian tract.

The title was *Wings Over Britain*
and the little crosses

turned out to be spitfires
of the Free Polish Squadron.

In a Polish Forest

A father walking his little son
between two columns of silver birch

looks immemorial. They're absorbed
in the national pastime

with the courtly shade of Pan Tadeusz
when there was leisure

to name the seven kinds of mushroom
and roll their syllables lovingly on the tongue.

But they disappear and it's like
a pit dug across the path

the forest changes name
to *Waldbezirk 77*

light thickens and dogs
are barking in a ring

at those *goings on*
in grey sealed lorries

ripping in and out
with a tearing of gears.

Sonderbehandlung, like a new
fertilizer, he joked

to the passing forester
who saw him, the jolly policeman,

that weekend,
mushroom picking with his family.

All Souls

There were trams all day
to the new industrial cemetery
crowded under a factory
between two motorways

and everywhere the word *rodzina*,
family this and family that.
The youngest dead had the most flowers,
heaped chrysanthemums in plastic coats

fringing the oval photograph.
Families picnicked on deckchairs
among the dear departed,
it might have been a fishing outing,

or a flower show on marble.
Candlesmoke poured
from little windproof jars.
After dark, they make a blaze.

*   *   *

Someone knew of a hole in the wire.
We wriggled through,
into the pitted acres
and walked by feeble torchlight

fearing for our ankles
in a blackness rarified
by occasional gravelights
which showed the extent,

a huge field of smashed slabs
and coarse grass.
Dense, intricate language
fractured under the hammer.

There were small hands for a child,
the ark and books, the broken tree,
the crown, and a flower for Myriam,
the first decipherable name.

Below the perimeter wall
we found the unmarked pits
intended for those who dug them
—but they ran out of time—

and on the wall, crowded
with recent writing, a misprint:
in memory of these,
this plague has been erected. . .

*   *   *

The Chairman *has commended the construction*
of an excellent road to the cemetery . . .
The building of exemplary roads under present conditions
is a monument to the ghetto's vitality.

The ghetto *Chronicle* for 1941.
There was a further 'Heartening Development':
Even today, garden plots have been turned over,
prepared, and are producing vegetables.

Overnight, an urban people
became greenfingered,
scraping and tilling
even the smallest scrap;

nothing went to waste. And if things
went according to plan, soon there would be
no ghetto at all, but one huge leisure park,
announced in the *Litzmannstadter Zeitung.*

The Satrap

'The strictest autonomy governs the ghetto; no German would dare touch my prerogatives, and I shall never let any do so.'

Mordechai Chaim Rumkowski, Eldest of the Jews in the Łódź ghetto, speaking in 1942

His milk and honey tongue
preaches work, peace, order,
care for the sick,
care for the young—

his silver head in Zion
his feet in the ghetto;
but all they cry
is *bread and potatoes*,

his famished workforce
of tailors, carpenters,
rag sorters, dyers,
gravediggers, archivists . . .

He keeps a bodyguard
of bullies and spies,
a mint, a bank, a press,
dominions, titles, powers,

a kingdom of soup kitchens;
and he cries to his people
I come like a bandit
for your children;

those I love above all!
(He is broken like them),
with a quota to be met
—and creditors grow cruel,

so he bows and scrapes,
and begs on his knees,
and weeps, and obeys . . .
His surprised Masters

can't believe their luck:
the children and the sick
are theirs, in mute processions,
regular as clockwork.

A Letter from America

Your letter came from Florida
where you were employed
to produce poets
in the lavish country
of Wallace Stevens.

Your springy students
bounced to ghettoblasters;
they owned convertibles
and spent the afternoon
in surfwear on the beach.

You had a magic haircut
achieved without scissors
and set off the smoke-alarm
by scorching the toast.
Once, in grubby London,

you said you admired
prosperity, your mind
on the German miracle.
Now, in your endearing
practical way

you ask if I have running water
and wonder, with a touch,
perhaps, of envy,
if life is 'abrasive'
in my part of the world.

Well, I did have running water;
when I pulled the bath plug
it ran all over the floor.
I drank vodka,
I queued,

I survived
'my heroic year'
as a Polish colleague put it
with glinting irony.
As a little girl

she lived without knowing it
alongside a people
in a corner of her city
who were encouraged
by Mr Wosk's *Sensational Ideas*

to utilize spoiled potatoes,
to produce ice,
to avoid using glass
in the windows,
and to make the most of coal dust.

The Heroes

(Sw. Brygidy Church, Gdańsk)

The heroes
are already in bronze.
They journeyed briefly
from their cradles
and their godfearing mothers
to immortality
on a reredos.

The world was simpler
when freedom
was a dockyard
draped in the red and white
of bloodied sheets
hung from windows
after the massacre.

Certitude
was the scarred Madonna
gazing out
from someone's lapel,
the sound of boots
striking an iron bridge
and gates clanging to.

AUTHOR'S NOTE

For some of the historical detail in Part 3 I am indebted to two remarkable books, *The Chronicle of the Łódź Ghetto*, edited by Lucian Dobroszycki (Yale, 1984), and *Łódź Ghetto*, edited by Alan Adelson and Robert Lapides (Viking, 1989).

Łódź, (pronounced 'Woodge'), is Poland's second city with a population of some 850,000. It is situated 120 kms south-west of Warsaw. From the time of its rapid growth in the nineteenth century, its chief industry has been textiles. Łódź was renamed Litzmannstadt by the Nazis at the time of its integration into the Wartheland late in 1939.

Both 'The Satrap' and 'All Souls' make reference to Mordechai Chaim Rumkowski, chosen by the Nazis as official Eldest of the Jews in the Łódź ghetto. He was usually addressed as the 'Chairman'. Rumkowski's ardent vision of the ghetto as a Jewish State in miniature was accompanied by an extraordinary mixture of energetic megalomania and paternalistic compassion. Labouring under the delusion that by hard work and good behaviour the Jews of Łódź might survive the war, he became an ideal servant to his Nazi masters, satisfying their deportation quotas (beginning with the children and the sick) in the hope that each deportation might be the last. Along with the vast majority of ghetto dwellers, Rumkowski was himself deported with his family to Auschwitz, where he perished in the summer of 1944. He remains a figure 'drenched in duplicity', to borrow Primo Levi's phrase; but given the appalling circumstances in which he had to make choices, Levi adds the caveat: 'We are all mirrored in Rumkowski, his ambiguity is ours . . .'

OXFORD POETS

Fleur Adcock
Kamau Brathwaite
Joseph Brodsky
Basil Bunting
Daniela Crăsnaru
W. H. Davies
Michael Donaghy
Keith Douglas
D. J. Enright
Roy Fisher
David Gascoyne
Ivor Gurney
David Harsent
Gwen Harwood
Anthony Hecht
Zbigniew Herbert
Thomas Kinsella
Brad Leithauser
Derek Mahon

Jamie McKendrick
Sean O'Brien
Peter Porter
Craig Raine
Henry Reed
Christopher Reid
Stephen Romer
Carole Satyamurti
Peter Scupham
Jo Shapcott
Penelope Shuttle
Anne Stevenson
George Szirtes
Grete Tartler
Edward Thomas
Charles Tomlinson
Chris Wallace-Crabbe
Hugo Williams